21st Century Disasters

E-Waste in Guiyu, China

by Shannon Berg

FOCUS READERS
BEACON

www.focusreaders.com

Focus Readers is distributed by North Star Editions:
sales@northstareditions.com | 888-417-0195

Produced for Focus Readers by Red Line Editorial.

Photographs ©: Imaginechina/AP Images, cover, 1, 7; Norman Ng/KRT/Newscom, 4, 14; Red Line Editorial, 9; Damrong Rattanapong/Shutterstock Images, 10–11; Poravute Siriphiroon/Shutterstock Images, 13, 29; Eugene Hoshiko/AP Images, 16; Jim Xu/Getty Images News/Getty Images, 19; Tyrone Siu/Reuters/Newscom, 21; VILevi/Shutterstock Images, 22–23; Aly Song/Reuters/Newscom, 24; baranozdemir/iStockphoto, 26

Library of Congress Cataloging-in-Publication Data
Names: Berg, Shannon, author.
Title: E-waste in Guiyu, China / by Shannon Berg.
Description: Lake Elmo, MN : Focus Readers, [2020] | Series: 21st century disasters | Audience: Grades 4-6. | Includes index.
Identifiers: LCCN 2019006120 (print) | LCCN 2019011963 (ebook) | ISBN 9781641859455 (pdf) | ISBN 9781641858762 (hosted ebk.) | ISBN 9781641857383 (hardcover) | ISBN 9781641858076 (pbk.)
Subjects: LCSH: Electronic waste--China--Guiyu Zhen--Juvenile literature. | Salvage (Waste, etc.)--China--Guiyu Zhen--Juvenile literature. | Recycling (Waste, etc.)--China--Guiyu Zhen--Juvenile literature.
Classification: LCC TD799.85 (ebook) | LCC TD799.85 .B46 2020 (print) | DDC 363.72/88--dc23
LC record available at https://lccn.loc.gov/2019006120

Printed in the United States of America
Mankato, MN
May, 2019

About the Author

Shannon Berg is a young adult novelist, short story and nonfiction author, and poet. Shannon lives in Minnesota with her husband, two children, dog, and cat. Most of her spare time is spent working her way through the stack of books waiting to be read on her nightstand.

Table of Contents

Chapter 1

A Growing Problem

Thick smoke filled the air in Guiyu, China. The city's rivers were black and full of trash. Piles of used and broken **electronics** covered the ground. This type of garbage is known as e-waste.

Workers collect metal scraps from piles of trash in Guiyu, China.

Guiyu is near the coast of the South China Sea. In the 1990s, ships began bringing tons of e-waste to Guiyu. Trash came from all over China. Other countries sent their e-waste to Guiyu, too.

People in Guiyu worked to **recycle** the e-waste. They took apart the electronics. Then they sold the pieces. This work was dangerous. It released many chemicals. Some chemicals made people sick. Others ruined the land and water.

Women in Guiyu take apart cell phones.

E-waste poured into Guiyu from around the world. The piles of trash grew. So did the **pollution**. By 2013, Guiyu was the largest e-waste site on Earth.

China began making changes. One law changed how e-waste was recycled. People could no longer work on their own. Instead, waste had to go to an official **facility**. China also stopped accepting e-waste from other countries.

Even so, e-waste remained a problem. Guiyu still held tons of

Did You Know?

At one time, 70 percent of the world's e-waste was sent to China.

E-WASTE TIMELINE

1995
People begin recycling e-waste in Guiyu.

2001
China imports 1.7 million tons (1.5 million metric tons) of e-waste.

2007
As many as 100,000 workers recycle e-waste in Guiyu, often in small workshops.

2013
Guiyu becomes the largest area of e-waste in the world.

2015
China's government moves the workshops to a recycling facility.

2018
New laws in China ban e-waste from other countries.

garbage. Broken pieces spilled from houses and lined the streets. And ships kept bringing e-waste from other parts of China.

CASSETTE
Quick Charger
CHARGE

Chapter 2

Understanding E-Waste

E-waste is any trash that has electronic parts. Most e-waste comes from old or broken devices. TVs, computers, and cell phones are common sources. But e-waste comes from other **appliances**, too.

E-waste includes keyboards, phones, computers, and many other devices.

Each time people buy a new device, they must do something with the old one. They may throw it away. Or they may recycle it. Either way, the device becomes e-waste.

Compared to other trash, e-waste is hard to recycle. Devices are often made from several materials. For example, a device might use metal,

Did You Know?

In 2016, people created nearly 100 billion pounds (45 billion kg) of e-waste.

Cell phones often have glass screens, metal circuitry, and plastic cases.

plastic, and glass. These materials are often blended together.

Recycling facilities can separate the materials safely. But most people in Guiyu could not. Some took apart trash in their homes. Others worked in the city's streets.

Workers take apart computers on Guiyu's streets.

To separate the materials, people often melted or burned the pieces. Sometimes they washed the pieces in **acid**.

These actions put people in danger from chemicals. Many

devices contain toxic metals, such as mercury or lead. Touching these metals can make people sick. Burning them causes pollution.

Some chemicals make the air unsafe to breathe. Others get into the water or ground. Plants cannot grow. People and animals get sick.

Did You Know?

Old flat-screen TVs can release mercury if they're burned or thrown away. People who breathe this metal get sick.

Chapter 3

Life in Guiyu

In the late 1990s, families in Guiyu began opening small **workshops**. They recycled e-waste to earn money. They sold the pieces and materials. Whole families worked together. Even the kids helped.

Workers take apart computers to collect the valuable parts inside.

Yang Linxuan's family owned one of these workshops. They melted phones and computers to get metal. Many families on their street did the same thing. Smoke from the burning trash filled the air. It made Yang want to cough.

Dr. Xia Huo studied the pollution in Guiyu. She found that many children had too much lead in their blood. This metal makes it hard for people to learn and grow. The pollution caused other problems,

Burning e-waste releases smoke and chemicals.

too. Chemicals in the air burned people's skin and lungs.

In 2013, China's government made a new law. It said e-waste could be recycled only at a facility.

The workshops had to move or close. Leaders hoped this change would **reduce** pollution.

Chen Xinrong owned a workshop for many years. After China's law changed, he had to move. Chen now makes less money. And he has less e-waste to recycle. But Chen supports the changes. Like many

Did You Know?

By 2012, Guiyu had more than 5,000 recycling workshops.

China's new law meant workshops like this one had to close.

people in Guiyu, he wants the city to be less polluted.

China's new rule seemed to help. By 2015, the air and water contained less metal than they did in 2014. The streets in Guiyu became cleaner as well.

E-Waste

Everyone can work to create less e-waste. People can choose to buy fewer electronics. Or they can buy new devices less often. Some stores will take back old items. They send the parts to be recycled.

Other changes help people recycle e-waste safely. Safe facilities use a special **process** to take items apart. Workers clean the pieces as they go. They keep the parts in containers. The containers prevent spills of oil or other fluids. Companies that make devices can help as well. They can look for ways to use fewer toxins.

Clean facilities help workers recycle e-waste safely.

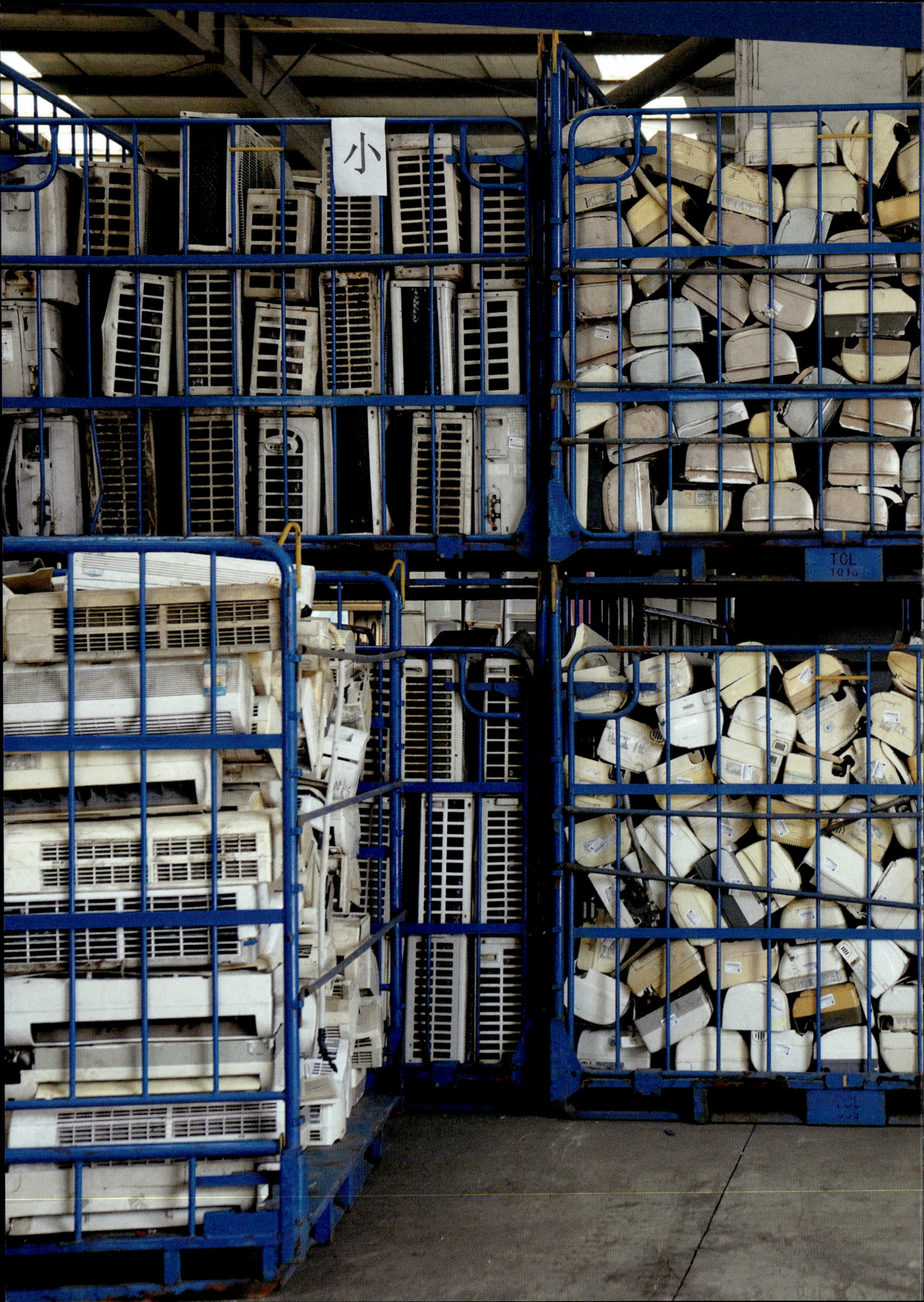

小
TCL

Chapter 4

Next Steps

Guiyu still recycles e-waste. The waste comes from all over China. It goes to recycling facilities. These buildings are designed to be clean and safe. They use filters to catch dust so it will not harm workers.

Air conditioners wait to be recycled at one of Guiyu's recycling facilities.

A worker wears gloves and a mask for protection while taking apart a computer monitor.

Safe facilities also clean the water they use. That way, they won't pollute rivers and lakes. Even so, dangerous chemicals are released. To stay safe, some workers wear gloves and masks.

Pollution in Guiyu has begun to improve. The air and rivers are cleaner. But e-waste remains a problem around the world. Many countries still create huge amounts of e-waste each year. Unless people make more changes, this problem will continue to grow.

Did You Know?

In 2018, China created more e-waste than any other country. The United States came in second.

FOCUS ON

E-Waste in Guiyu, China

Write your answers on a separate piece of paper.

1. Write a sentence describing one problem caused by e-waste.
2. Do you think the problem of e-waste can be solved? Why or why not?
3. At first, where did most e-waste recycling in Guiyu take place?
   A. in the city's rivers and lakes
   B. in official facilities
   C. in small workshops or the city's streets
4. Why do workers at recycling facilities wear gloves and masks?
   A. so other workers don't know who they are
   B. so they don't have to work as hard
   C. so they don't touch or breathe dangerous chemicals

5. What does **materials** mean in this book?

*Devices are often made from several **materials**. For example, a device might use metal, plastic, and glass.*

A. noises made by machines
B. supplies used to make items
C. devices with large screens

6. What does **toxic** mean in this book?

*Many devices contain **toxic** metals, such as mercury or lead. Touching these metals can make people sick.*

A. harmful or poisonous
B. helpful and healthy
C. soft and fuzzy

Answer key on page 32.

Glossary

acid
A strong chemical that can break down things placed in it.

appliances
Machines that have specific jobs, such as stoves or fans.

electronics
Machines that use electricity to work.

facility
A building or place with a specific use or purpose.

pollution
Harmful substances that collect in the air, water, or soil.

process
Steps people follow to create a result.

recycle
To take apart an old item and make the pieces into something new.

reduce
To lower the size, amount, or weight of something.

workshops
Places where people do jobs that involve making or fixing things.

To Learn More

BOOKS

Flynn, Sarah Wassner. *This Book Stinks!* Washington, DC: National Geographic, 2017.

LaPierre, Yvette. *Providing Waste Solutions for a City*. Minneapolis: Abdo Publishing, 2019.

Pratt, Mary K. *How Can We Reduce Household Waste?* Minneapolis: Lerner Publications, 2016.

NOTE TO EDUCATORS

Visit **www.focusreaders.com** to find lesson plans, activities, links, and other resources related to this title.

Index

Answer Key: 1. Answers will vary; **2.** Answers will vary; **3.** C; **4.** C; **5.** B; **6.** A